I0393359

Copyright

Through-Wall Motion Detection Using GPR

Copyright © 2016, 2017 by Dennis J Johnson

Author: Dennis J Johnson

First Printing February 2017

Version 1.3

ISBN 978-1542747578

EAN 1542747570

Disclaimer: The information and surveys identified in this book are intended to be generic.

This book is dedicated to the employees of Geophysical Survey Systems Inc. and UltraVision Security Systems LLC. Their dedication to the product and industry is unsurpassed.

Cover photo, first test of LifeLocator, back cover photo is LifeLocator PDA display

Contents

Introduction

Ground Penetrating Radar, GPR

The book focuses on GPR and is aimed at the evolution of a new use for the technology: detecting motion through concrete walls. It is written so that lay people, not from the world of geophysical technologies, can better understand the use of Ground Penetrating Radar (GPR) to detect and track moving targets and stationary breathing targets – behind walls and through debris.

The material presented demonstrates the product development and effectiveness of GPR technology to help accomplish the challenging objective of finding hidden and buried people. The photos contained in this book highlight just a sample of the varied testing and uses of GPR in the new field of motion detection. Since this application of GPR is very new to the world of security and search and rescue (SAR) there will clearly be many more discoveries and future applications for this technology

*   *   *

Author Dennis Johnson is the past president of Geophysical Survey Systems Inc.(GSSI), and past president of UltraVision Security Systems, Inc (UVSS). GSSI is the original pioneering GPR company and world leader in the development and manufacture of GPR systems used to detect and display buried objects. Both companies are owned by Oyo Group, Tokyo, Japan. The author was the product line champion and inventor on several patents related to using GPR for motion detection.

In the development of motion detection, GSSI designed and manufactured all sensors. The later spin-off division of UVSS developed systems software and carried out all marketing, sales, installation and training efforts.

The author has personally used GPR to survey many of the below ground uses of GPR. These surveys are outlined in two other books; one focused on archaeology the other on golf course greens and athletic fields.

Overview of the Technology

Ground Penetrating Radar (GPR) is one of several technologies that can be used to 'look' underground – and now through walls. Remember that every technology has its advantages and disadvantages. GPR technology has been in commercial use for over 40 years to identify subsurface features in applications ranging from quality assurance on bridge decks, highway infrastructure and construction projects to locating underground utilities, assessing snow and ice thickness, aiding forensics at mass gravesites and pinpointing features for archeological digs. See a more complete list of standard GPR applications in the back of this book.

Ground Penetrating Radar is a subset of 'standard' commercial aircraft radar that works in the time domain. However GPR uses a wide spectrum transmitter (Ultra-Wide Band) and transmits at lower frequencies. The total power emitted by a GPR transmitter is 1% of a cell phone power transmission. This makes the motion detection sensors extremely difficult to detect.

Turning science fact into entertainment fiction, standard GPR systems have been featured in popular TV shows such as *CSI* and *Law and Order: SVU*. Characters in books by best-selling authors Kathy Reichs, Steve Berry, John Sandford and Dan Brown have used GPR to help solve crimes. The 2016 mini-series "Mars" features GPR being used to identify potential landing sites for the first manned mission.

GPR has also gone off-planet for real. China's first moon mission, launched in December 2013, has GPR loaded onto the unmanned lunar rover to investigate the lunar crust and to map and measure the structure and depth of the lunar soil. The near-future ExoMars Rover mission will contain a GPR payload designed not only to map but also to collect and analyze samples of WISDOM (Water/Ice and Subsurface Deposit Information on Mars).

On a more down-to-earth level, GPR is being used to efficiently examine large surface areas and identify subsurface infrastructure and large objects bearing further investigation (e.g., foundations of buried cities and buildings, gravesites, water wells, old roads).

Data collection grids are used to collect the data. Much can be seen in real-time (2-D view) during data collection. Each survey line is saved as a computer file. Survey lines from the area being investigated are combined on a computer to create a 3-D volume that can be sliced and diced vertically and horizontally to understand the layers and anomalies underground or in walls.

In many technologies the sensor must exhibit the opposite of what is detected. Examples might be; cameras that must be dark to capture light, thermal sensors that must be cold to capture heat, etc. The significant change from GPR underground surveys morphing to GPR motion detection is that the antenna moves to collect static data in ground surveys and the antenna must be held stationary to detect motion.

GPR engineers have worked hard over the years to block out any operator motions during survey data collection. Survey operators are told to remove metal objects from their person to avoid having a potential reflector that might show up in the data. So blocking out motion was part of the engineering assignment.

The break through in the development of motion detection came when it was decided to find new applications and markets for GPR technology. GSSI needed a new path for growth.

For illustrative purposes computer screen captures display much of the data in this book. The data is more impressive when viewed directly on a computer in real time. Various color tables can and are used in the display to highlight moving targets.

Early Tests and Developments

This section describes some of the proof-of-concept tests conducted from 2003 through 2005. These findings drove the development of initial products and commercialization of market applications.

Wall or barrier where sensor is located

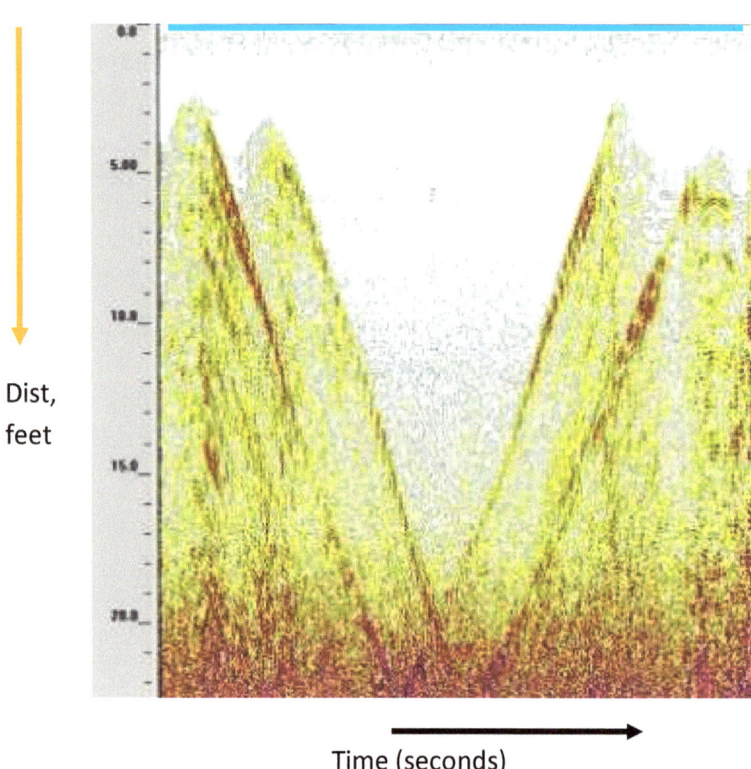

Dist, feet

Time (seconds)

The above is a screen capture from a very early test (early 2003) of motion detection using a SIR-3000 GPR system and a 400 MHz antenna. Both the SIR-3000 and antenna were standard production items using standard software. The sensor detection was made through a concrete block wall shown here as a blue line at the top of the screen. The yellow arrow on the Y axis indicates target distance from that wall.

This test shows two people briefly walking together toward the wall (two feet), walking away from the wall (out to twenty-two feet) and then walking back toward the wall. Target speed can be calculated from the slope of the lines. This screen configuration with the X axis as 'time' and the Y axis as 'distance' from the sensor is standard for all motion tests shown in this book.

A SIR-3000 with a 400 MHz antenna looking through a test wall of reinforced concrete. The equipment used was off-the-shelf.

This is one of many pictures taken from the first public test of the system to detect people in debris piles. This day long test in August 2003 took place at the Fairfax County Search and Rescue (SAR) training site in Virginia. Note the SIR-3000 was used here with cable attached to the 400 MHz antenna. The hardware and software were all standard production items. Only the software setup was changed from normal GPR survey use.

The final test of the day was to find a test subject (fireman) who hid in the debris pile. The subject was found in less than 15 minutes. The complete test report by a department of Homeland Security can be found referenced in the back of this book..

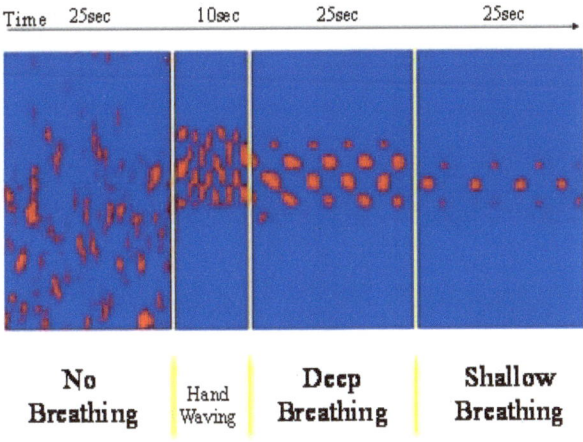

Identification of a man breathing behind a concrete block wall

This is a screen capture from a very early test (2004) of motion detection using a SIR-3000 GPR system and a 400 MHz antenna. Again both the SIR-3000 and antenna were standard production items. The antenna 'looked' through a 6 inch reinforced concrete slab.

The target was a stationary man making the various motions indicated in the four modes. The man was about 4 feet behind the concrete slab.

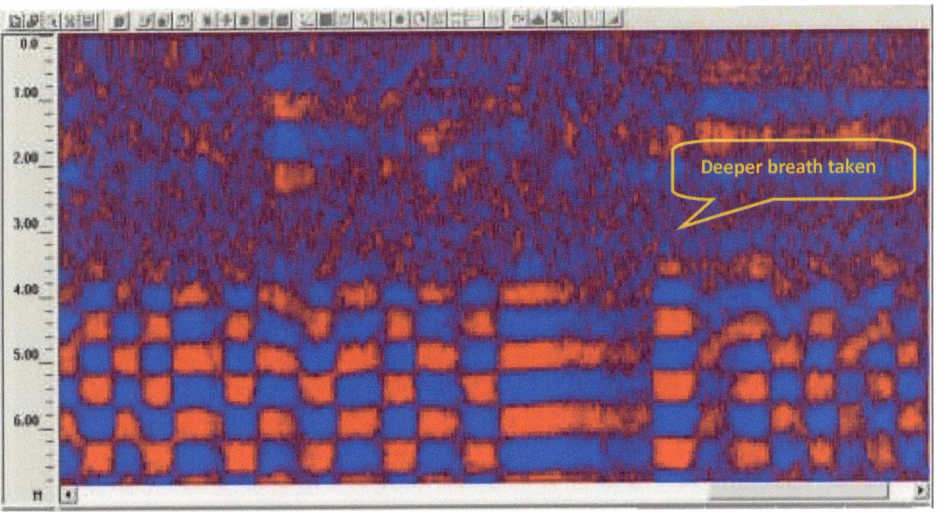

This is another breathing test where the subject is 4 feet behind a concrete wall. The subject can be seen breathing normally on the left, he is asked to hold his breath; he stops breathing about 7 seconds and then takes a deeper breath before resuming normal breathing.

Based on these seemingly successful initial through-wall tests and many others made over the course of six months, various other test configurations using buried sensors (antennas) were devised.

In this test setup the sensor was buried under the road passing by the old GSSI building in the spring of 2004. A 400 MHz antenna was used to conduct various motion tests of moving targets on the road. The antenna was covered with 6 inches of dirt and asphalt.

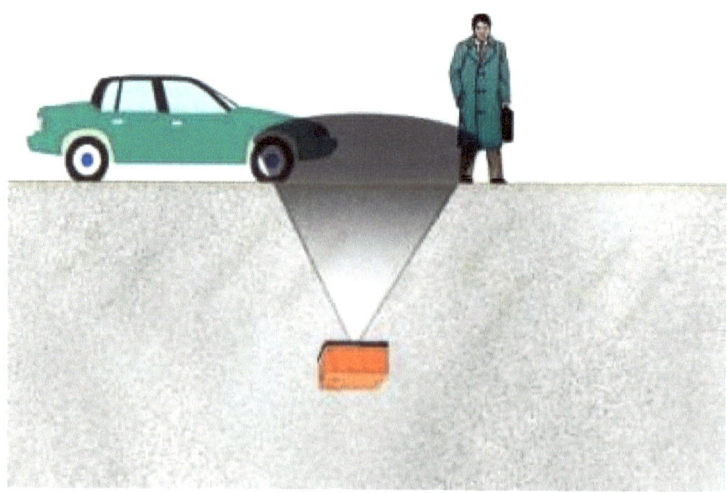

This drawing shows the detection concept with the road sensor.

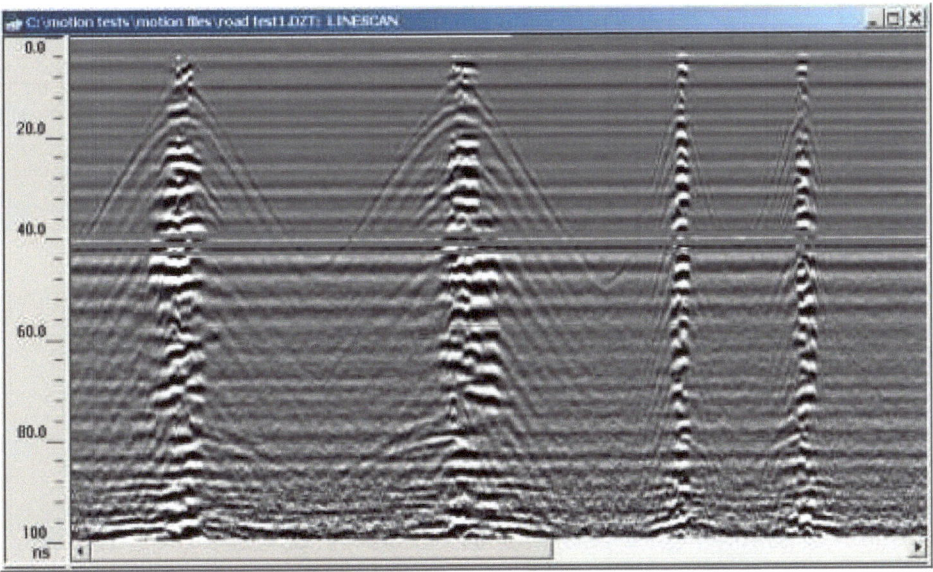

A man target walked across the road sensor in both directions and then ran across the sensor in both directions. The hyperbola in the walking tests shows the distance to the man approaching and then departing. Note the slope of the hyperbola determines the target speed.

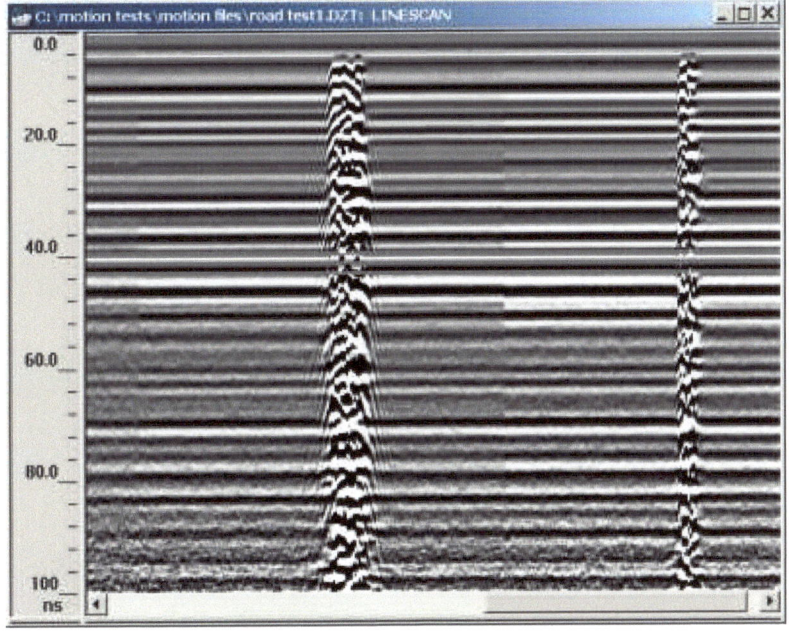

A Jeep passed over the road sensor at 5 mph and then 20 mph. Note that target distance is measured in time, nanoseconds, in these tests. The GPR system used was off-the-shelf.

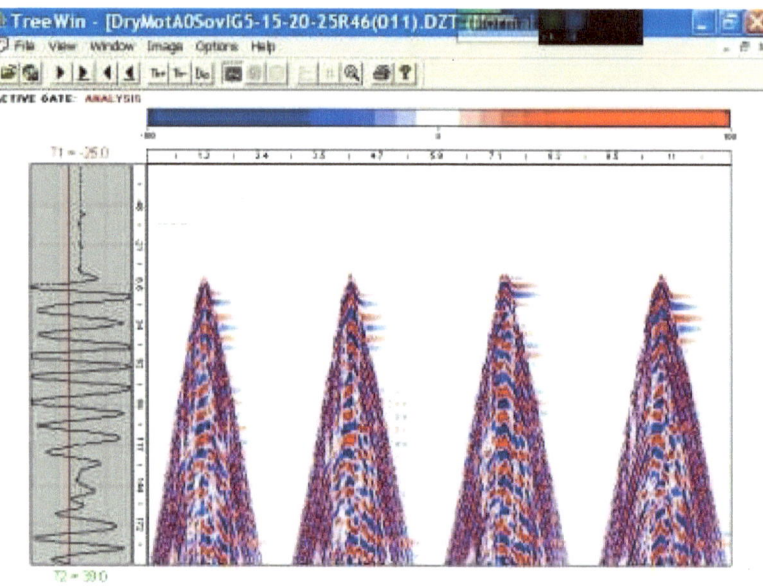

In the spring of 2005 a Midshipman at the US Naval Academy took on a term paper project to evaluate different sensor setups for through-wall motion detection. Dr. Tony Mucciardi was the course instructor. This is just one of the screen captures where the subject was walking a race track pattern. Detection was made through a concrete block wall and the display above shows four circuits made by the test subject as he approached the wall and walked away 4 times.

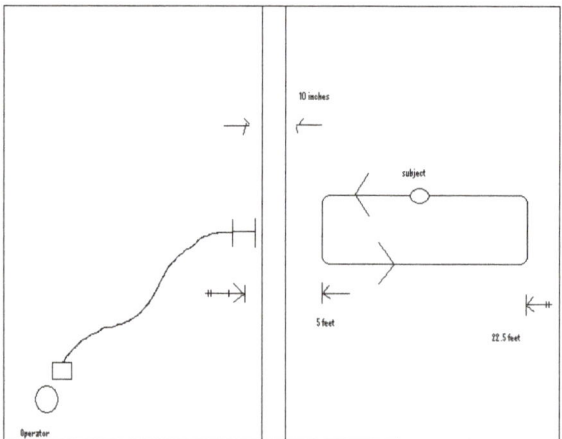

This diagram taken from the term paper shows the system layout and test subject's counter-clockwise path. The double vertical line in the middle represents the concrete block wall. Both pictures here were taken from Midshipman Joel D. Dunivant's report. The report made several very useful observations regarding antenna orientation.

LifeLocator

After various tests and the critique by the Fairfax SAR unit, the engineering department at GSSI worked on a revised product with two primary objectives: removal of the cable connection to the antenna and development of a custom PDA to make the display device smaller and easier to read. This new product was now given the name 'LifeLocator"

This and the next picture show subsequent tests in early 2006 at Fairfax SAR in Virginia using the new LifeLocator. These follow-on tests used the much improved display software which required minimal operator training to detect buried victims, either moving or just breathing.

This is one of many motion and breathing tests conducted at the FEMA debris pile near Beverly, Massachusetts. This debris pile was normally used to train SAR personnel and search dogs. The test pile was built primarily from destroyed concrete bridge deck materials. In this case the ability of the system to detect *breathing* of the man in the red helmet through 5 feet of bridge deck concrete and rebar was impressive.

The FEMA people at this location were a great help in aiding with tests and ongoing product development. The tests at this location were conducted over a 2 year period, 2006 and 2007.

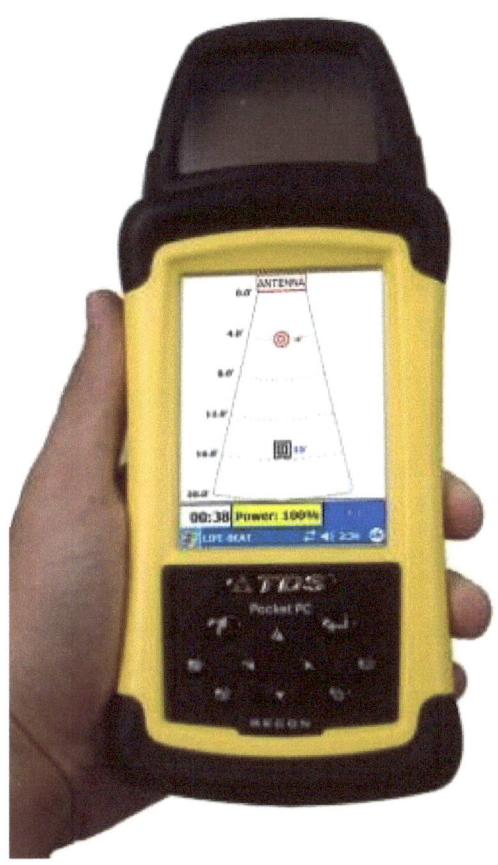

This is the second (and final) PDA that was developed for the first product sales of LifeLocator. The display shows a person breathing (red circle) and a person moving (black square) in a debris pile. The distance, or depth, to each person can clearly be seen. The size of the red circle and black square increases with increased detection confidence.

Note that at a given distance the 'victim' can be located anywhere within a 120 degree detection arc or cone. For most debris pile excavation and rescue this degree of detection capability is adequate. More accurate target detection requires moving the sensor to a second location, or the use of a second sensor.

This is one of many tests made by the Tokyo Fire Department before they purchased LifeLocator units. Notice the breathing 'target' inside the concrete pipe. Photo from Oyo Group.

LifeLocator Field Use

A devastating earthquake hit China May 10, 2008. By this time almost 25 LifeLocators have been sold to ten different Fire Departments in China. The earthquake area became the first large scale test and proving ground for the system. The following pictures show Chinese fire department personnel and troops being mobilized for transport to the earthquake zone.

LifeLocator units and troops boarded large cargo airplanes that took them to the devastated areas. One of the UVSS Chinese sales representatives was previously trained in SAR.

This is one of many sites where the LifeLocator was put to work finding buried victims who were still alive. It is worthwhile to note that LifeLocator finds people ONLY if they are moving or breathing. Slow breathing as in a coma is also detectable.

Two more locations show the LifeLocator in use. The lower photo shows a crew digging to find a detected victim. *Over 49 documented people were saved by the LifeLocator* during the first 4 days of rescue operations. One person was detected and rescued after being buried over 70 hours. The China International SAR team worked with various city fire departments covering the wide affected area. All photos from China are from sales employees at Ancom, Beijing.

19

Ten days after the earthquake and back in Beijing, a TV news crew made the LifeLocator famous for 15 minutes. These photos were taken at the Ancom sales office.

This is a scanned picture of an Israeli newspaper article showing members of the Israel International SAR team in Haiti. This was the second major field test of LifeLocator. They are searching for buried victims after the earthquake in 2010. Both French SAR members and Israeli SAR members saved over 50 buried people using LifeLocator. The French team saved one person who was buried over 3 days.

It is worthwhile to note that the LifeLocator was tested early in its development by Israel SAR leaders. A number of good suggestions and input regarding search methodology came from these people. Part of the search process was to make a red paint mark on the spot where 'life' was detected and a black paint mark on the surveyed spot where no life was detected. The following excavation teams could quickly understand where to dig and, equally important, where not to dig.

The LifeLocator quickly sold in 12 countries (but not in the US). As a portable SAR device there was no comparable equipment on the market.

Product Line Expansion

UltraSensor Security

It quickly became clear from the previously documented road testing that buried sensors also had a place in the market. It was time to find other uses for the technology.

The UltraSensor System was developed to improve facility security where the system could be administered and monitored from anywhere in the world. A new server with new operating system software was developed.

One significant advantage of the UltraSensor System is that it extends and broadens the sercurity perimeter of existing systems in a very covert manner. Now instead of waiting for a window or door to be breached, security forces can detect people as they approach the building.

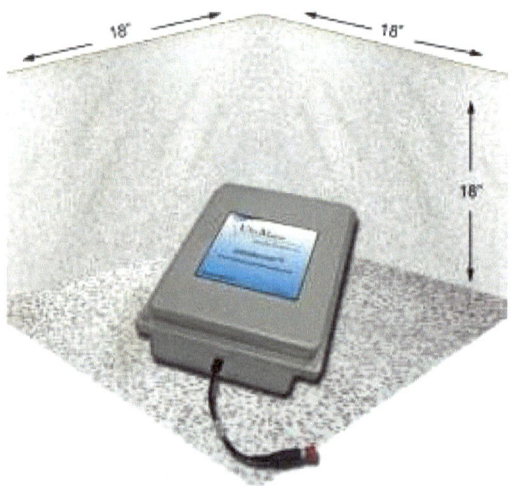 

Two views of the UltraSensor antenna
showing burial requirements

Buried sensors cannot be seen or detected. Sensors were connected to a control box through a CAT 5 cable Power Over Ethernet (POE). PTZ cameras were an integral part of each system installation. The system was designed to easily integrate with existing security systems.

This concept drawing shows the detection volume of a single buried sensor. Hemisphere diameter is 50 feet. Not a perfect detection circle on the ground, but very close.

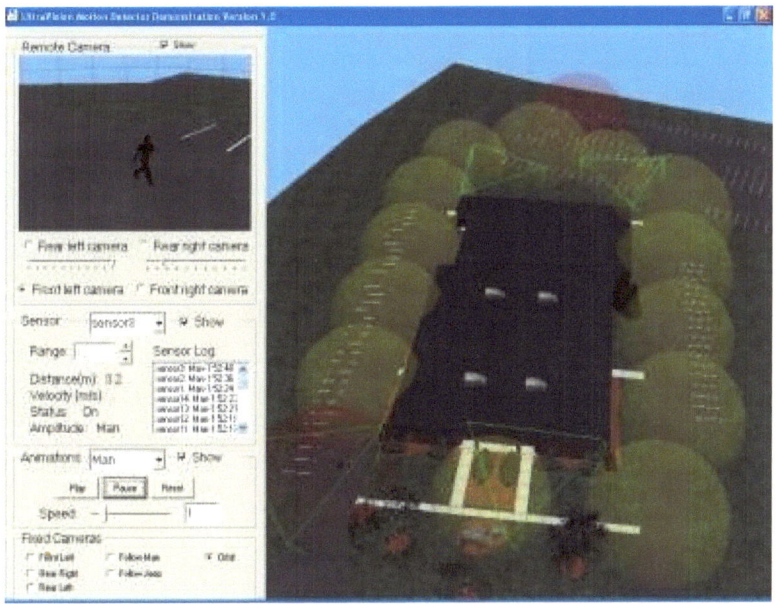

Computer simulation of the UltraSensor security system installed at the GSSI facility in Salem, NH. As the animated car or person circled the building the PTZ cameras on each corner of the building would automatically follow the target. The upper left camera view is following an animated person. The actual system was installed at the facility several months later.

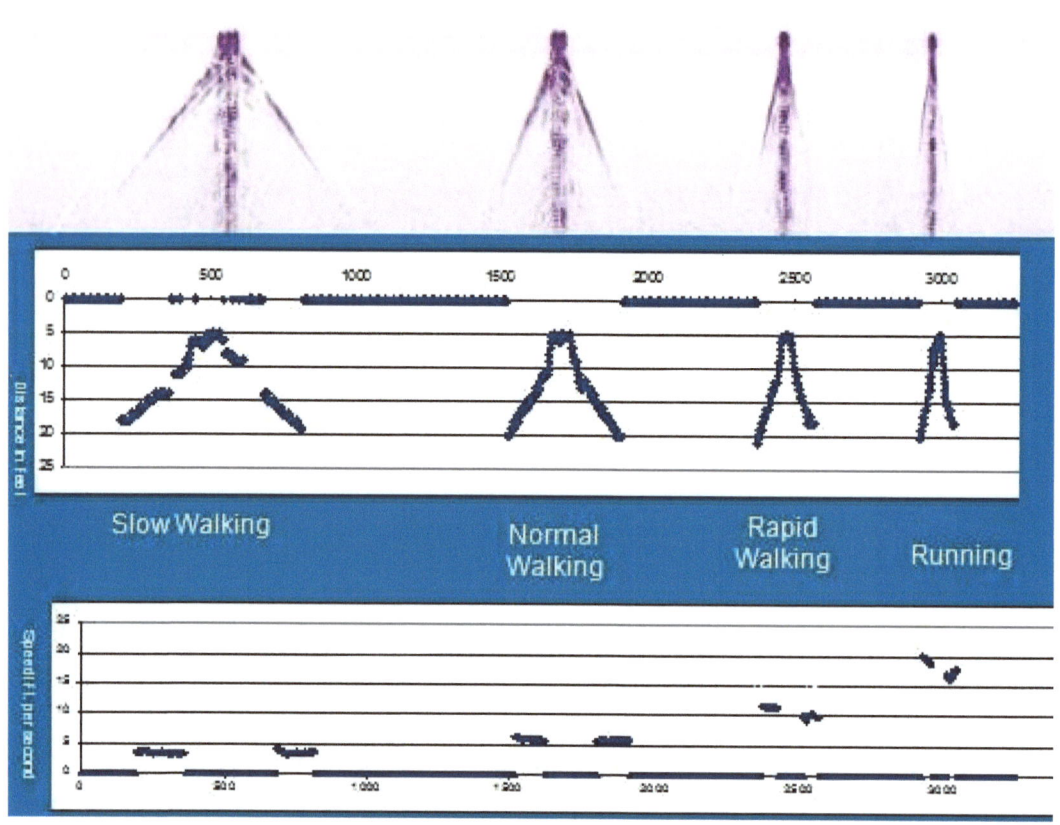

The upper graph shows detection distances, the lower graph shows target speed.

This screen capture shows the development of UltraSensor software. The raw data shown on page 11 is now converted to a more readable format with a calculated velocity. Detection distances shown here are 20 feet from the sensor. Initial production software increased this distance to just over 30 feet (radius).

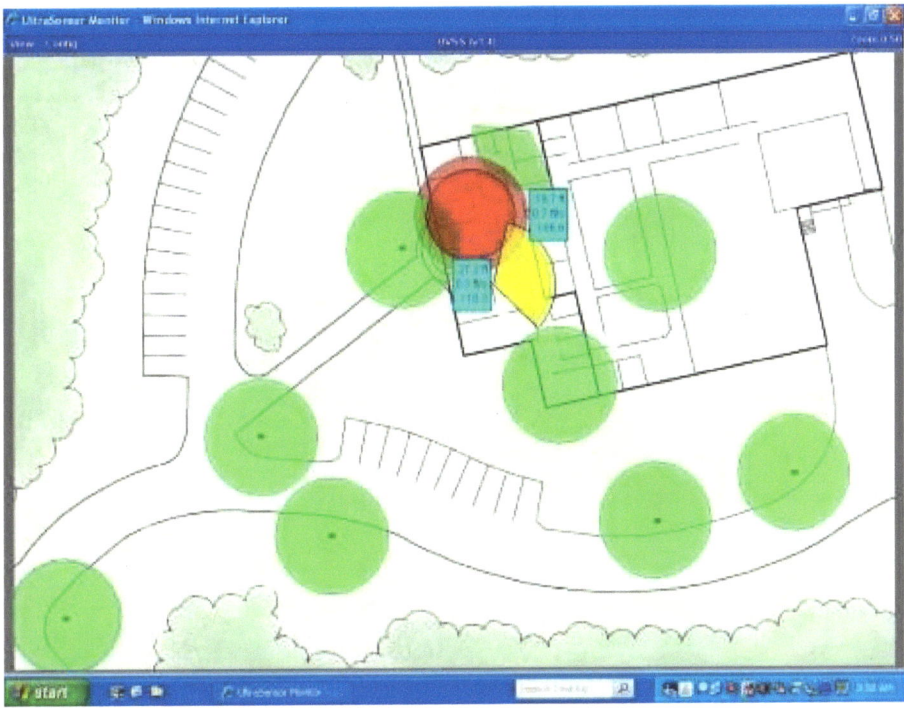

This is a screen capture of the UltraSensor security system installed and working at the UVSS facility in New Hampshire (2008). A scaled diagram of the facility and grounds was made and sensors were placed on the diagram in their actual locations. Eleven sensors are installed here. Circles scaled to correct detection distances represent buried detectors looking upward. Arcs indicate sensors sitting on edge looking in one direction only. The sensor in the middle of the building was roof mounted. All sensors are buried or hidden! Notice how much the security perimeter is expanded beyond the building walls. Everyone entering the driveway is automatically recorded on cameras that track the target(s).

The green detection areas indicate no motion. The red detection area indicates the sensor detected an unauthorized person. The yellow sensor detection area indicates a person was detected but in an authorized area. Note that within the blue detection box; target distance, speed and target amplitude are all listed.

PTZ cameras were installed on each corner of the building. Anyone entering the detection zones would be tracked and automatically recorded.

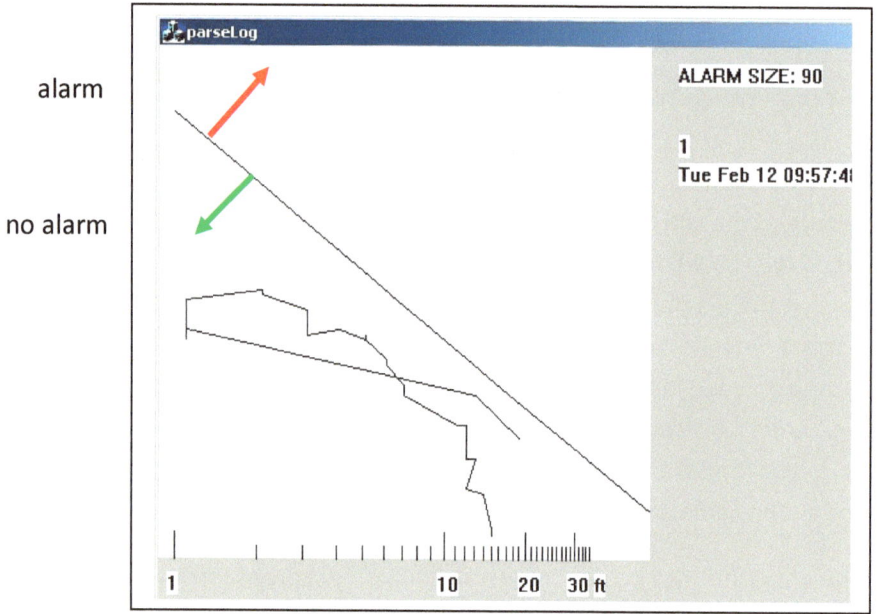

This development test shows the ability of the UltraSensor system to differentiate small moving targets from people. The sensor was buried in the ground with a pile of snow on top. The location is marked with the snow shovel (and yellow arrow) near the center of the picture. A small white dog is enticed to walk near the sensor. The test screen capture shows the return signal of the wandering dog compared to the alarm threshold. The return signal from a man sized target is above and right of the threshold shown.

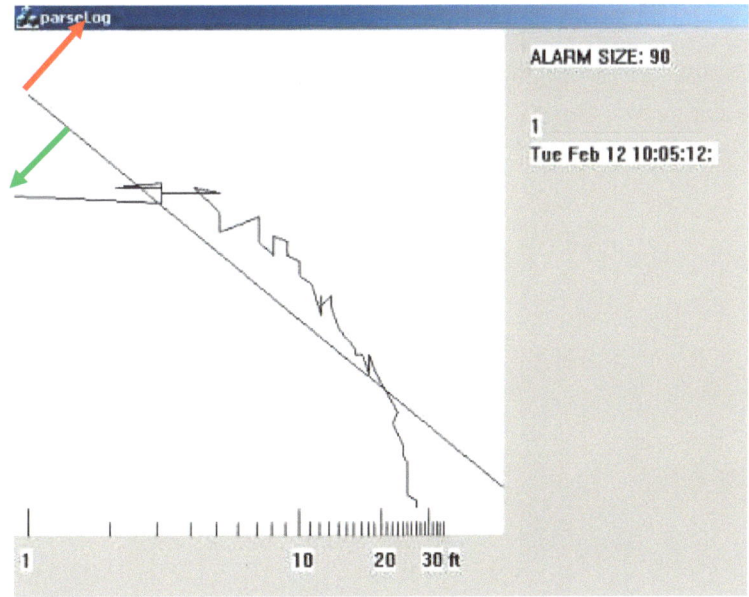

This is the same sensor test using a walking man as a target. The amplitude of the returned signal is sufficient to exceed the alarm threshold. Notice that in this beta software version the man was detected 21 feet from the sensor. The newest software version could detect people twenty-five feet from the sensor. The target speed makes no difference in detection capability. Rain and snow conditions are also below the detection threshold.

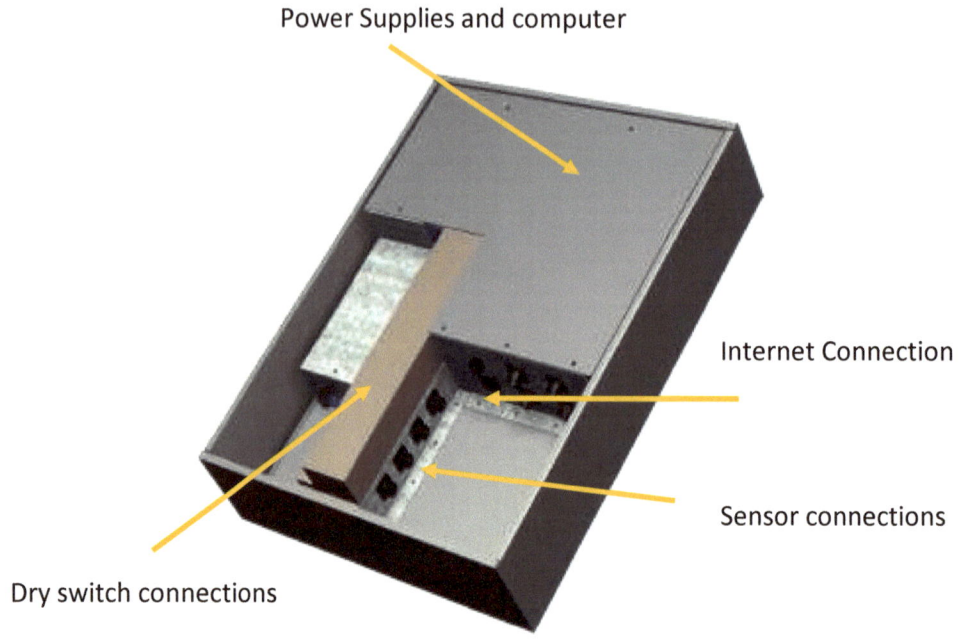

Power Supplies and computer

Internet Connection

Sensor connections

Dry switch connections

This is a 3D CAD model of an UltraSensor server. The unit is a wall-mounted metal box. Up to 24 sensors can be connected here. Servers came in three sizes based on the number of sensor connections required; 8, 24, 48.

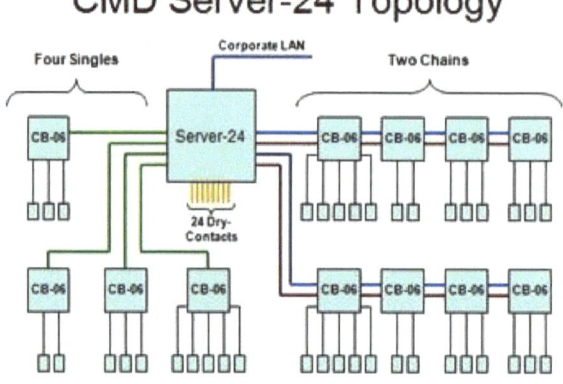

CMD Server-24 Topology

UltraSensor systems were installed in museums in the US and Europe, high end residential estates in the US, Ireland, England and Portugal and in high-value warehouses. PTZ cameras were an integral part of each system.

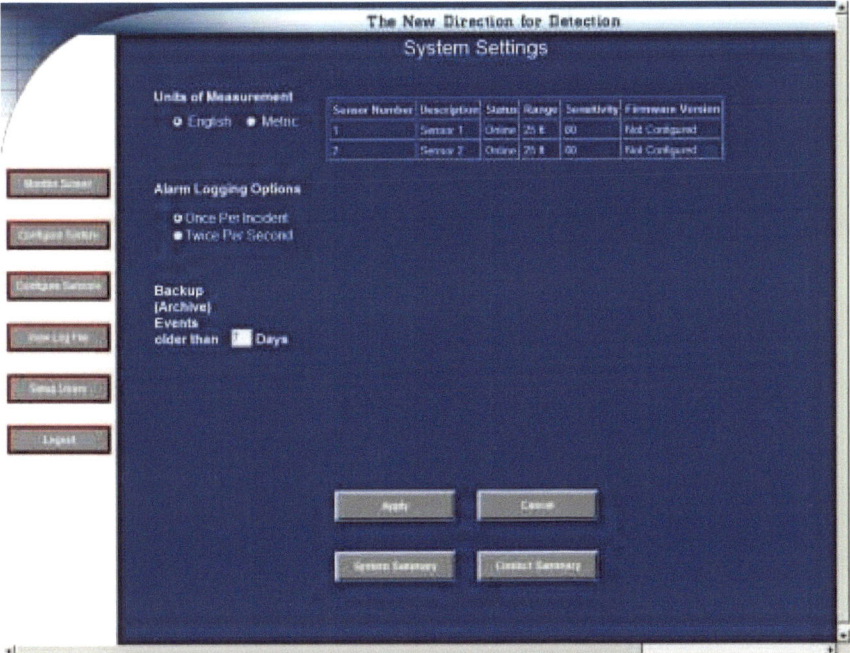

There are 4 set up screens. This one shows the amplitude sensitivity setup of two sensors

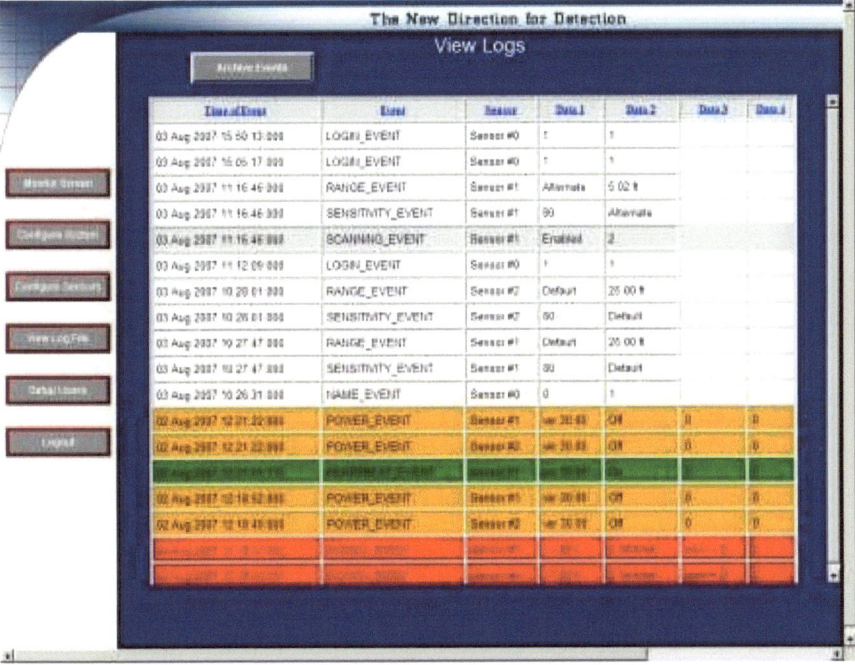

This is the event logging screen. All system changes and all alarms are logged for later analysis if necessary. All data is automatically archived after seven days.

SwatVision

The next logical step in advancing the product line was to introduce units that could be used in police and military operations. Fundamentally LifeLocator was turned to look horizontally instead of vertically downward. Software was developed so a user could monitor multiple sensors simultaneously. A new rugged housing was designed. The product name became 'SwatVision'.

This picture shows two different sensors developed for SwatVision, a 400 MHz antenna on the left and a 270 MHz antenna on the right. Both units could be connected via Wi-Fi or radio frequency to a computer laptop. Up to 5 sensors could be monitored simultaneously on the laptop.

In one demonstration of the system the 400 MHz antenna was able to track a moving person through 2 feet of heavily reinforced concrete.

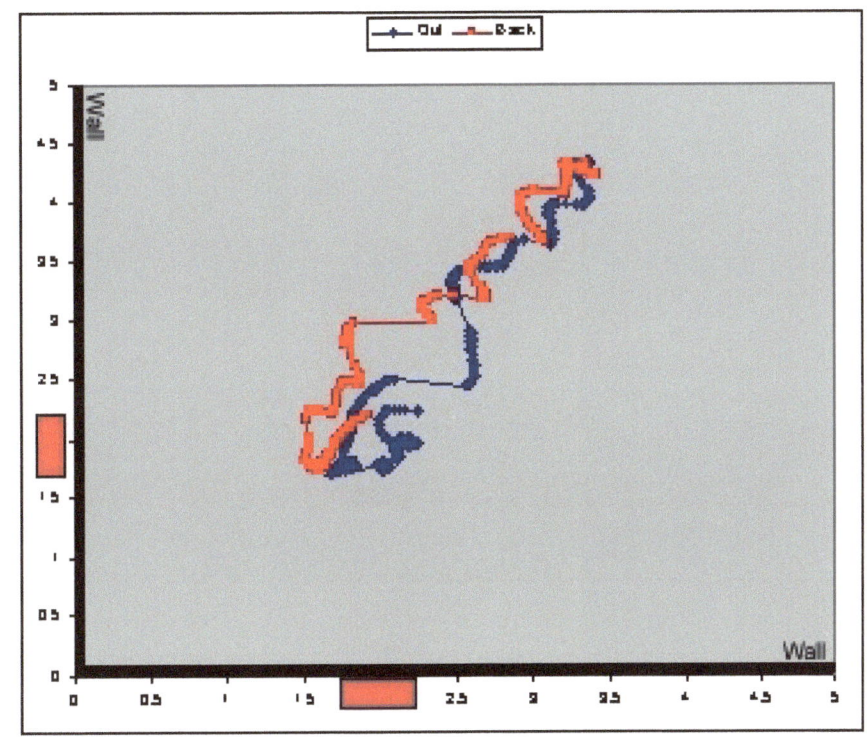

This is a screen capture showing an early prototype software test (2003) that later led to SwatVision. Two sensors were placed outside the walls and a walking person was tracked. The orange boxes represent the location of each sensor. The blue trail is a person walking away from the sensors and the red trail is a person walking toward the sensors. This test shows the ability to not only get the location and speed of the target but a good sense of target direction (negative or positive velocity).

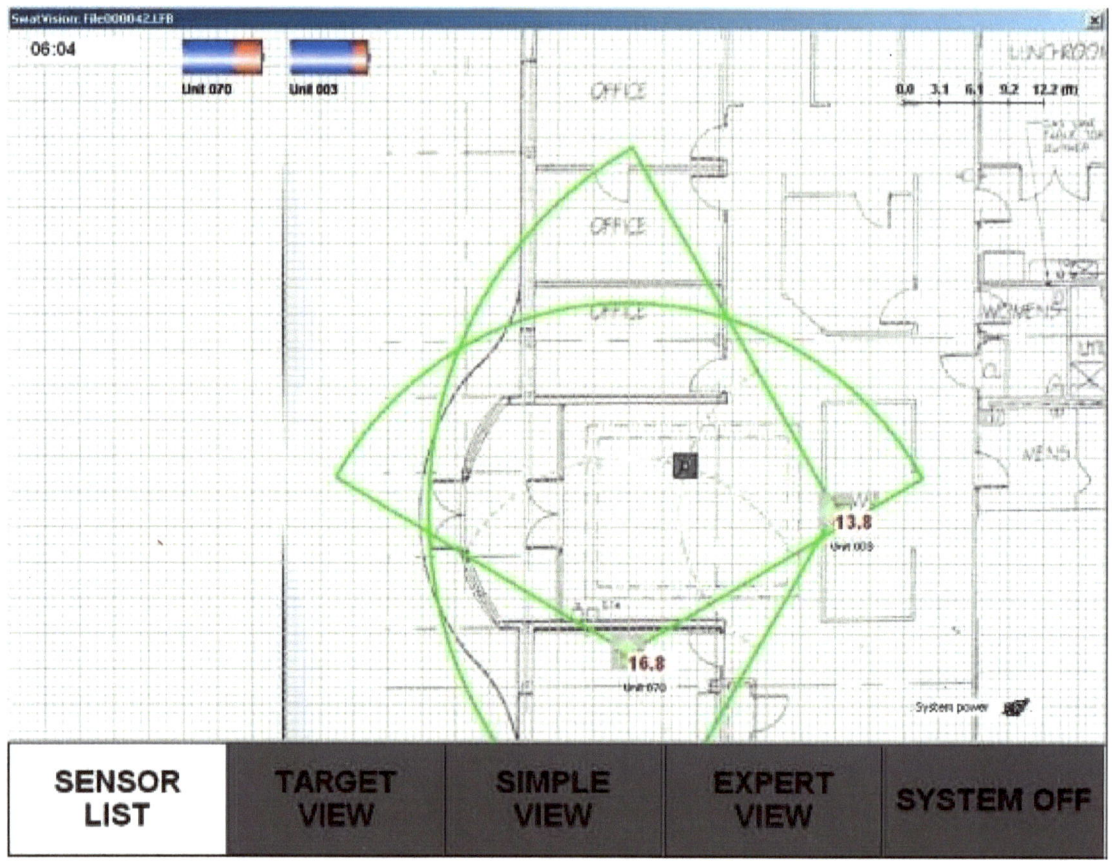

This screen capture shows the final version of SwatVision in action. A building layout was imported and used to position the sensor locations. The display scale was user selectable based on the imported building plan. Here two sensors are shown with the green arc indicating the area of detection. Where the arcs overlap a targets position can be shown. The sensors were hidden behind a stone wall and a wood wall. The black square indicates the position of a person walking through the building. The moving person's location display is shown within a one second delay from actual position.

Like LifeLocator the SwatVision software display could also show a stationary person with a red circle e.g. the software can easily detect breathing of a motionless person. Note the sensor battery display, upper left, and the distance to the target is displayed at each sensor location.

Again our Israeli sales representatives and members of their military conducted prototype tests and helped define system specifications and use.

This is another screen capture from the SwatVision control laptop. The utility program updates and changes sensor software. All changes are logged for quality control.

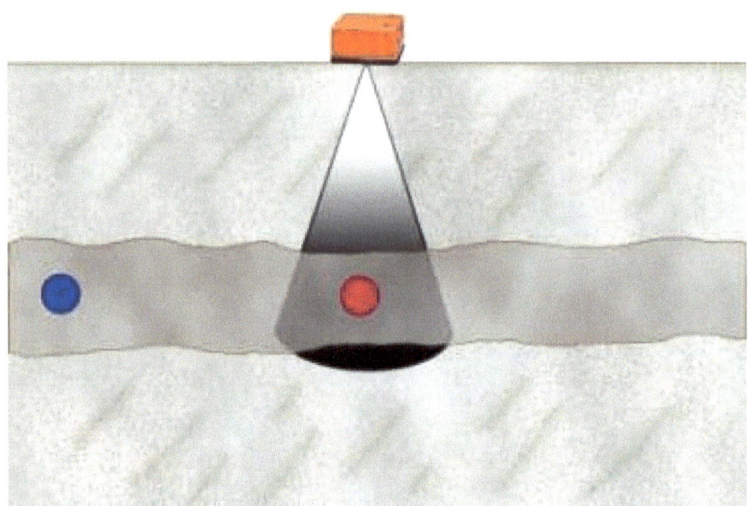

This concept diagram shows another use of both SwatVision and UltraSensor. The sensor, used portable or fixed, can be placed to track people moving through tunnels or sewer pipes. The sensor is clandestine, possibly hidden behind concrete, and cannot be seen, detected or damaged by transiting people. One such system was tested in the El Paso, Texas area where people passing through a sewer line were counted automatically.

Summary

Through wall motion detection is now a reality. Three different end uses are described here showing the versatility and usefulness of the GPR technology.

Interestingly the same essential radar printed circuit board is used in all three applications. The primary product line difference lies in the system software and user interface.

Target detection range is only limited by the transmitter power. All three product lines here conform to FCC and CE emission regulations. The detection range for the LifeLocator working in a debris pile varies from 15 to 20 feet. The variation is due to differences in the composition of the debris pile. The range of SwatVision through one foot of reinforced concrete is about 35 feet. The longer range is due to the fact that the SwatVision detection transmission path is mostly air. Round trip signal attenuation is a smaller factor.

Detection distances can be greatly increased if higher power transmitters are used. Also continued improvements in software and antenna configurations will lead to ever better performing systems.

Visit the GSSI website, www.geophysical.com, to learn more about the continuing evolution of the technology and products.

Other GPR Surveys

There are many other CURRENT uses for GPR technology in addition to motion detection.

- **Concrete Structure Surveys**
  Coring, Cutting, Drilling concrete
- **Utility Pipe Detection and Mapping**
  - **Highway Inspection**
  - **Airport Runway Inspection**
  - **Bridge Inspection**
  - **Railroad Inspection**
    Prevent derailments
- **Golf Course and Athletic Field Analysis**
  - **Ice & Snow Thickness**
  - **Archaeology discoveries**
    - **Forensics**
  - **Tree Trunk and Root Surveys**
  - **Minerals/Vug Detection**
- **Architectural Façade inspection**
- **Pipe and Storage Tank Leaks**
  - **Tunnel Detection**
    - **Arms Caches**
    - **UXO Detection**
    - **Mass Graves**
- **Microphone Detection in Concrete Walls**

GPR Glossary

Antenna – The transmit/receive antenna (sensor) used in all of these products was 400 MHz and 270 MHz (Megahertz). The rule of thumb with UWB (Ultra Wide Band) radar is that the transmit signal is the same power from one-half this frequency to twice the frequency. In other words, a 400 MHz antenna transmits 'uniformly' from 200 MHz to 800 MHz.

Color table – Various colors may be applied to the GPR data to highlight reflected radar amplitudes. For example, the grayscale color table highlights positive return peaks as white and negative return peaks as black. A varying degree of gray is used for smaller signal amplitudes in between the high positives and low negatives. In other color tables, a blue-for water, green-for-turf color spectrum helps people relate better to GPR data.

GPR (Ground Penetrating Radar) – This form of electromagnetic transmitter and receiver may be compared to conventional radar systems that operate in the time domain. The primary difference is conventional radar systems operate at higher frequencies within a very narrow frequency spectrum, while GPR operates at lower frequencies and with a very broad transmission band of frequencies. This wide band transmission is known as 'Ultra Wide Band' (UWB).

PTZ Camera – Pan, Tilt, Zoom video cameras are a staple in the security industry. In the UltraSensor system these cameras are programmed to turn and focus on any pre designated area when an UltraSensor detects motion. The security guard simply has to look at the monitor; no operator action is required to guide cameras and record intruders.

Questions or Comments? Contact the author, Dennis Johnson at Below the Turf, LLC – 603-490-0922, and www.belowtheturf.com.

Product innovation awards:

- Frost and Sullivan - Innovative Security Product of the Year 2008
- Chesapeake Regional Tech Council - Innovator of the Year 2009 (Annapolis, Washington DC, Baltimore)
- ISC West Trade Show Las Vegas, Nevada - Security Product of the Year
- IFSEC Trade Show Birmingham, UK - Security Product of the Year
- New Hampshire High Tech Council - Product of the Year (Life Locator)

References

Operational Test and Evaluation Report TerraSIRch SIR 3000
Barrier Penetrating Radar for Locating Casualties in Rubble
Conducted on 11 August 2003
by the Virginia Task Force 1 Urban Search and Rescue Team
at the Fairfax County, VA Fire and Rescue Training Center
Testing managed by the Emergency Response Technology (ERT) Program

UltraVision
Security Systems, Inc.

The New Direction for Detection

US Exec Meets with Successful Chinese Rescue Team

New York, June 19, 2008 - During a visit last week to the Peoples Republic of China to review the performance of the firm's LifeLocator® Search and Rescue System, UltraVision Vice-President Bill Lozon met with several regional rescue teams from Chengdu, Chongqing, Beijing, the China International Search and Rescue Squad and the Hebei Fire Brigade. During the visit, Mr. Lozon was able to document several instances where LifeLocator saved lives including one victim buried for over 70 hours!

Fire officials from the Chongqing Fire Brigade, the Beijing Rescue Squad, and Hongbo Si, Manager of the China International Search and Rescue Organization (CISAR) all of which played a role in the rescue effort also lauded LifeLocator's ability to save valuable time by identifying search areas where no living trapped victims were detected thereby allowing rescuers to focus efforts on other areas while reducing risk of injury of the rescuers themselves.

Seismological Disasters & Chinese Emergency Response Mechanism

Dr. Sheo Pandey

C3S Paper No. 250 - Feb. 17, 2009

http://www.c3sindia.org/china-internal/481

China May 2008 Update

The high profile 184 member National Earthquake Team, which had arrived in Chengdu (Peoples Republic of China) at 22:40 CST on 12 May 08, carried out rescue operations in 48 places in 15 days. It included worst hit Dujiangyan, Hanwang town in Mianzhu, Yingxiu town in Sichuan and Beichuan. The team consisted of both soldiers and medics.

The team used UltraVision LifeLocator (TM) Life Detection System, capable of detecting life from the movement created from even shallow breathing. The team rescued 49 survivors, which included 6 experts trapped in the course of rescue operations, 30 students buried under the debris of collapsed school buildings and 13 residents of partially damaged and/ or fully razed to ground individual houses. Besides, the team cleared 1080 remains. It helped other teams in locating 12 survivors and rescuing 36 trapped residents.

Much of the rescue operation was literally handled by 146000 PLA and PAPF troops and 75000 militia and reservists. Ge Zhenfeng, deputy chief of the general staff, and Wang Guanzhong, director of the CMC General Office, reportedly took initial command of the situation, ordering units from the Air Force, Chengdu Military Region, Jinan Military Region and the Armed Police Force, as well as airborne unit to quickly proceed to the stricken areas to rescue quake victims.

www.ingramcontent.com/pod-product-compliance
Lightning Source LLC
Chambersburg PA
CBHW041318180526
45172CB00004B/1141